Animal Groups

Birds

By Nick Rebman

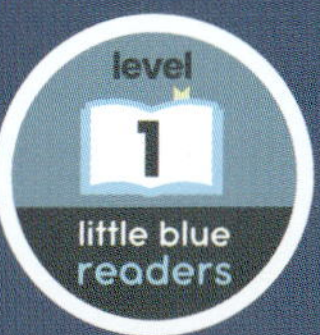

www.littlebluehousebooks.com

Little Blue House is distributed by North Star Editions:
sales@northstareditions.com | 888-417-0195

Produced for Little Blue House by Red Line Editorial.

Photographs ©: Shutterstock Images, cover, 4, 7, 9, 11, 13, 15, 17, 19, 21, 22–23, 24 (top left), 24 (top right), 24 (bottom left), 24 (bottom right)

Library of Congress Control Number: 2022919930

ISBN
978-1-64619-807-8 (hardcover)
978-1-64619-836-8 (paperback)
978-1-64619-893-1 (ebook pdf)
978-1-64619-865-8 (hosted ebook)

Printed in the United States of America
Mankato, MN
082023

About the Author

Nick Rebman is a writer and editor who lives in Minnesota. He enjoys reading, drawing, and going for long walks with his dog.

Table of Contents

wing

Birds

I see a bird.

It has wings.

I see a bird.

It has a beak.

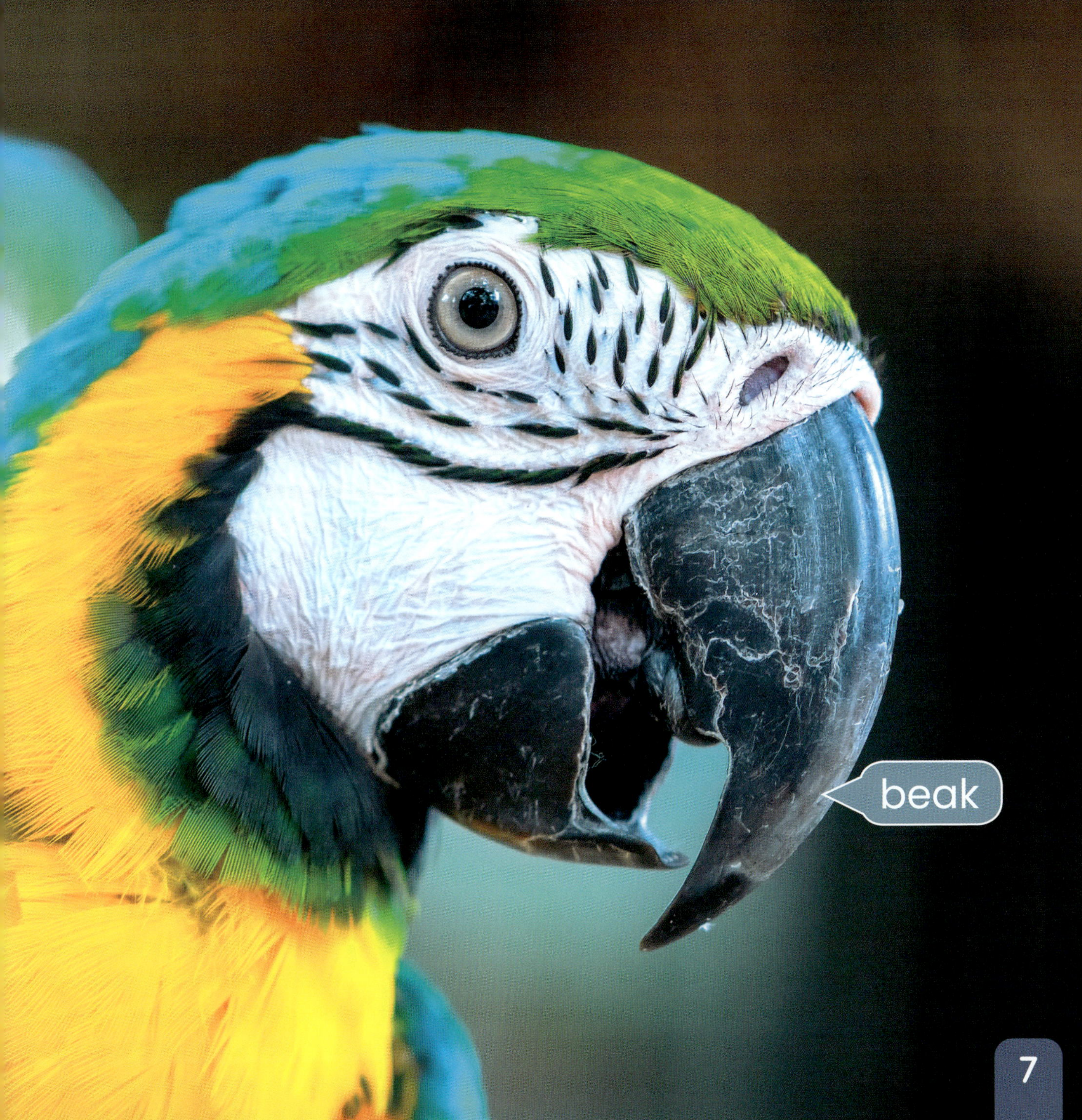
beak

I see a bird.

It has a tail.

tail

I see a bird.

It has feathers.

feather

I see a bird.

It has legs.

leg

I see a bird.

It has eyes.

eye

I see a bird.

It has a nest.

nest

I see a bird.

It has babies.

baby

I see a bird.

It has a worm.

worm

Is It a Bird?

All birds hatch from eggs.

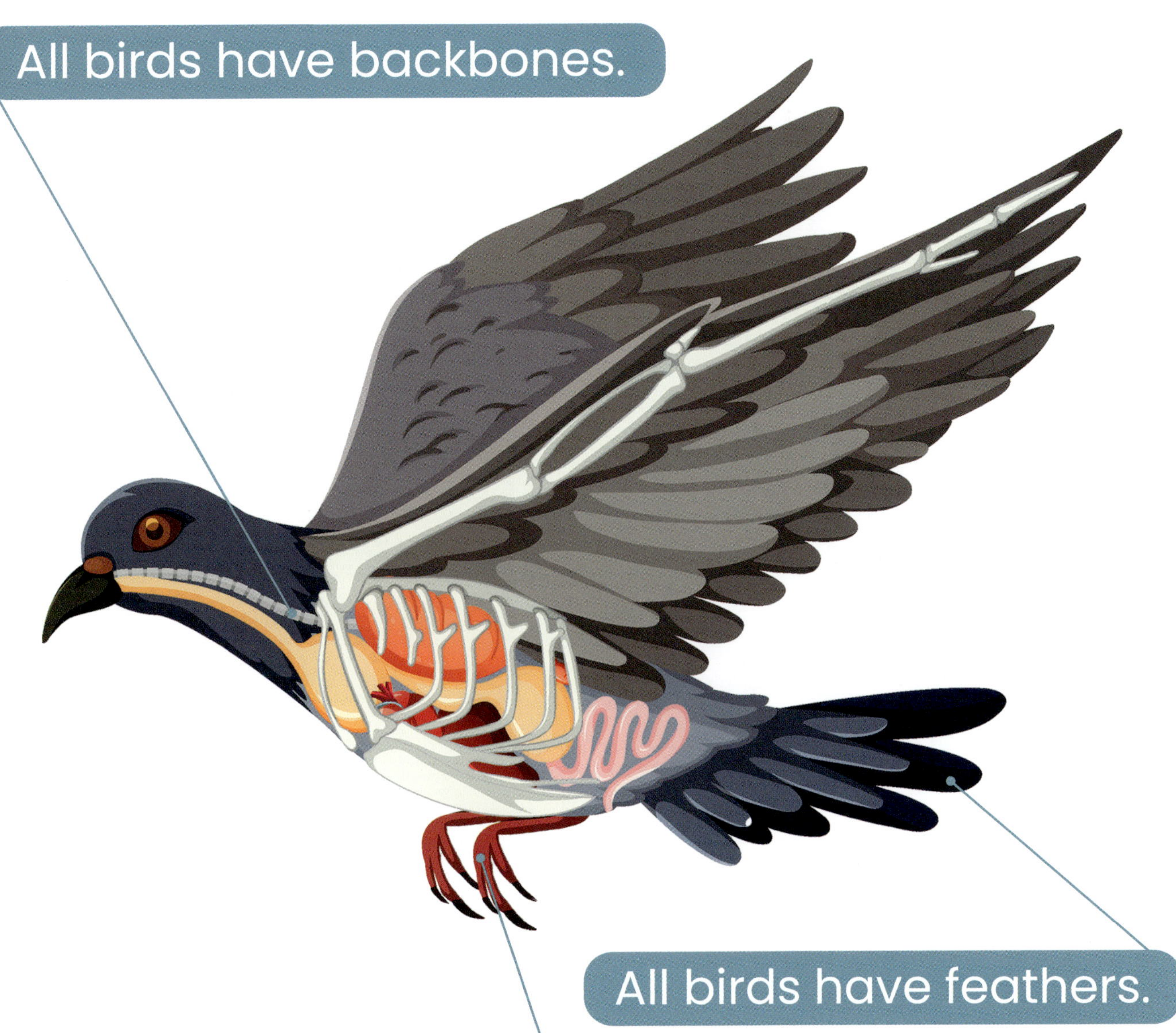
All birds have backbones.
All birds have feathers.
All birds have two legs.

Glossary

beak

nest

feathers

worm

Index